Higher
Mathematics

Leckie × Leckie

First exam published in 2005.
Published by Leckie & Leckie Ltd, 3rd Floor, 4 Queen Street, Edinburgh EH2 1JE
tel: 0131 220 6831 fax: 0131 225 9987 enquiries@leckieandleckie.co.uk www.leckieandleckie.co.uk

ISBN 978-1-84372-638-8

A CIP Catalogue record for this book is available from the British Library.

Leckie & Leckie is a division of Huveaux plc.

Leckie & Leckie is grateful to the copyright holders, as credited at the back of the book, for permission to use their material.
Every effort has been made to trace the copyright holders and to obtain their permission for the use of copyright material.
Leckie & Leckie will gladly receive information enabling them to rectify any error or omission in subsequent editions.

[BLANK PAGE]

X100/301

NATIONAL
QUALIFICATIONS
2005

FRIDAY, 20 MAY
9.00 AM – 10.10 AM

MATHEMATICS
HIGHER
Units 1, 2 and 3
Paper 1
(Non-calculator)

Read Carefully

1   **Calculators may <u>NOT</u> be used in this paper.**

2   Full credit will be given only where the solution contains appropriate working.

3   Answers obtained by readings from scale drawings will not receive any credit.

SCOTTISH
QUALIFICATIONS
AUTHORITY

FORMULAE LIST

Circle:

The equation $x^2 + y^2 + 2gx + 2fy + c = 0$ represents a circle centre $(-g, -f)$ and radius $\sqrt{g^2 + f^2 - c}$.

The equation $(x - a)^2 + (y - b)^2 = r^2$ represents a circle centre (a, b) and radius r.

Scalar Product: $\quad a.b = |a|\,|b|\cos\theta$, where θ is the angle between a and b

$$\text{or} \quad a.b = a_1 b_1 + a_2 b_2 + a_3 b_3 \text{ where } a = \begin{pmatrix} a_1 \\ a_2 \\ a_3 \end{pmatrix} \text{ and } b = \begin{pmatrix} b_1 \\ b_2 \\ b_3 \end{pmatrix}.$$

Trigonometric formulae:

$$\sin(A \pm B) = \sin A \cos B \pm \cos A \sin B$$
$$\cos(A \pm B) = \cos A \cos B \mp \sin A \sin B$$
$$\sin 2A = 2\sin A \cos A$$
$$\cos 2A = \cos^2 A - \sin^2 A$$
$$= 2\cos^2 A - 1$$
$$= 1 - 2\sin^2 A$$

Table of standard derivatives:

$f(x)$	$f'(x)$
$\sin ax$	$a\cos ax$
$\cos ax$	$-a\sin ax$

Table of standard integrals:

$f(x)$	$\int f(x)dx$
$\sin ax$	$-\dfrac{1}{a}\cos ax + C$
$\cos ax$	$\dfrac{1}{a}\sin ax + C$

ALL questions should be attempted.

Marks

1. Find the equation of the line ST, where T is the point (−2, 0) and angle STO is 60°.

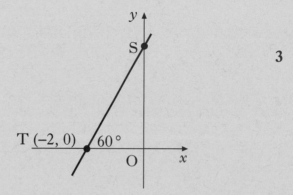

3

2. Two congruent circles, with centres A and B, touch at P.

Relative to suitable axes, their equations are

$x^2 + y^2 + 6x + 4y - 12 = 0$ and
$x^2 + y^2 - 6x - 12y + 20 = 0$.

(a) Find the coordinates of P.

(b) Find the length of AB.

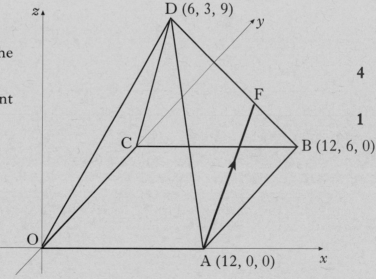

3

2

3. D,OABC is a pyramid. A is the point (12, 0, 0), B is (12, 6, 0) and D is (6, 3, 9).

F divides DB in the ratio 2:1.

(a) Find the coordinates of the point F.

(b) Express $\overrightarrow{AF}$ in component form.

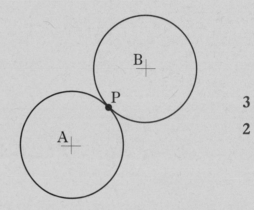

4

1

[Turn over

Marks

4. Functions $f(x) = 3x - 1$ and $g(x) = x^2 + 7$ are defined on the set of real numbers.

   (a) Find $h(x)$ where $h(x) = g(f(x))$.   2

   (b) (i) Write down the coordinates of the minimum turning point of $y = h(x)$.

        (ii) Hence state the range of the function h.   2

5. Differentiate $(1 + 2 \sin x)^4$ with respect to x.   2

6. (a) The terms of a sequence satisfy $u_{n+1} = ku_n + 5$. Find the value of k which produces a sequence with a limit of 4.   2

   (b) A sequence satisfies the recurrence relation $u_{n+1} = mu_n + 5$, $u_0 = 3$.

        (i) Express u_1 and u_2 in terms of m.

        (ii) Given that $u_2 = 7$, find the value of m which produces a sequence with no limit.   5

7. The function f is of the form $f(x) = \log_b (x - a)$. The graph of $y = f(x)$ is shown in the diagram.

   (a) Write down the values of a and b.   2

   (b) State the domain of f.   1

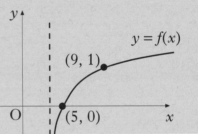

8. A function f is defined by the formula $f(x) = 2x^3 - 7x^2 + 9$ where x is a real number.

   (a) Show that $(x - 3)$ is a factor of $f(x)$, and hence factorise $f(x)$ fully.   5

   (b) Find the coordinates of the points where the curve with equation $y = f(x)$ crosses the x- and y-axes.   2

   (c) Find the greatest and least values of f in the interval $-2 \le x \le 2$.   5

9. If $\cos 2x = \dfrac{7}{25}$ and $0 < x < \dfrac{\pi}{2}$, find the exact values of $\cos x$ and $\sin x$.   4

Marks

10. (*a*) Express $\sin x - \sqrt{3} \cos x$ in the form $k \sin (x - a)$ where $k > 0$ and $0 \le a \le 2\pi$.   **4**

(*b*) Hence, or otherwise, sketch the curve with equation $y = 3 + \sin x - \sqrt{3} \cos x$ in the interval $0 \le x \le 2\pi$.   **5**

11. (*a*) A circle has centre $(t, 0)$, $t > 0$, and radius 2 units.

Write down the equation of the circle.   **1**

(*b*) Find the exact value of t such that the line $y = 2x$ is a tangent to the circle.   **5**

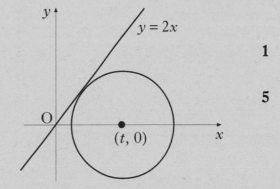

[END OF QUESTION PAPER]

[BLANK PAGE]

Official SQA Past Papers: Higher Maths 2005

X100/303

NATIONAL
QUALIFICATIONS
2005

FRIDAY, 20 MAY
10.30 AM – 12.00 NOON

MATHEMATICS
HIGHER
Units 1, 2 and 3
Paper 2

Read Carefully

1 **Calculators may be used in this paper.**

2 Full credit will be given only where the solution contains appropriate working.

3 Answers obtained by readings from scale drawings will not receive any credit.

FORMULAE LIST

Circle:

The equation $x^2 + y^2 + 2gx + 2fy + c = 0$ represents a circle centre $(-g, -f)$ and radius $\sqrt{g^2 + f^2 - c}$.

The equation $(x - a)^2 + (y - b)^2 = r^2$ represents a circle centre (a, b) and radius r.

Scalar Product: $\boldsymbol{a}.\boldsymbol{b} = |\boldsymbol{a}|\,|\boldsymbol{b}|\cos\theta$, where θ is the angle between $\boldsymbol{a}$ and $\boldsymbol{b}$

or $\boldsymbol{a}.\boldsymbol{b} = a_1 b_1 + a_2 b_2 + a_3 b_3$ where $\boldsymbol{a} = \begin{pmatrix} a_1 \\ a_2 \\ a_3 \end{pmatrix}$ and $\boldsymbol{b} = \begin{pmatrix} b_1 \\ b_2 \\ b_3 \end{pmatrix}$.

Trigonometric formulae:

$$\sin(A \pm B) = \sin A \cos B \pm \cos A \sin B$$
$$\cos(A \pm B) = \cos A \cos B \mp \sin A \sin B$$
$$\sin 2A = 2\sin A \cos A$$
$$\cos 2A = \cos^2 A - \sin^2 A$$
$$= 2\cos^2 A - 1$$
$$= 1 - 2\sin^2 A$$

Table of standard derivatives:

$f(x)$	$f'(x)$
$\sin ax$	$a\cos ax$
$\cos ax$	$-a\sin ax$

Table of standard integrals:

$f(x)$	$\int f(x)dx$
$\sin ax$	$-\dfrac{1}{a}\cos ax + C$
$\cos ax$	$\dfrac{1}{a}\sin ax + C$

ALL questions should be attempted.

Marks

1. Find $\int \dfrac{4x^3 - 1}{x^2}\, dx, \quad x \neq 0.$ **4**

2. Triangles ACD and BCD are right-angled at D with angles p and q and lengths as shown in the diagram.

    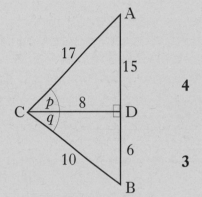

    (a) Show that the exact value of $\sin(p + q)$ is $\dfrac{84}{85}$.   **4**

    (b) Calculate the exact values of:

         (i) $\cos(p + q)$;

         (ii) $\tan(p + q)$.   **3**

3. (a) A chord joins the points A(1,0) and B(5,4) on the circle as shown in the diagram.

    Show that the equation of the perpendicular bisector of chord AB is $x + y = 5$.   **4**

    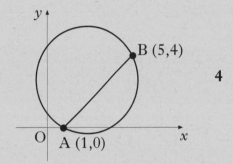

    (b) The point C is the centre of this circle. The tangent at the point A on the circle has equation $x + 3y = 1$.

    Find the equation of the radius CA.   **4**

    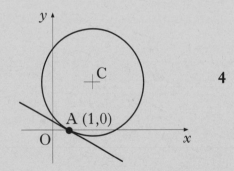

    (c) (i) Determine the coordinates of the point C.

         (ii) Find the equation of the circle.   **4**

    [Turn over

Marks

4. The sketch shows the positions of Andrew(A), Bob(B) and Tracy(T) on three hill-tops.

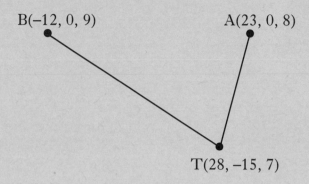

B(−12, 0, 9)   A(23, 0, 8)

T(28, −15, 7)

Relative to a suitable origin, the coordinates (in hundreds of metres) of the three people are A(23, 0, 8), B(−12, 0, 9) and T(28, −15, 7).

In the dark, Andrew and Bob locate Tracy using heat-seeking beams.

(a) Express the vectors $\overrightarrow{TA}$ and $\overrightarrow{TB}$ in component form.

2

(b) Calculate the angle between these two beams.

5

5. The curves with equations $y = x^2$ and $y = 2x^2 - 9$ intersect at K and L as shown.

Calculate the area enclosed between the curves.

8

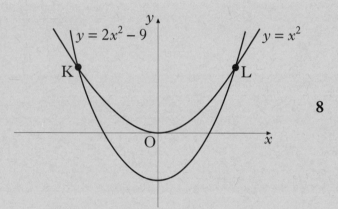

6. The diagram shows the graph of $y = \dfrac{24}{\sqrt{x}}$, $x > 0$.

Find the equation of the tangent at P, where $x = 4$.

6

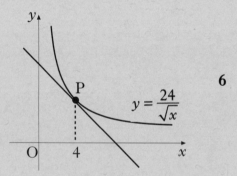

7. Solve the equation $\log_4(5 - x) - \log_4(3 - x) = 2$, $x < 3$.

4

Marks

8. Two functions, f and g, are defined by $f(x) = k\sin 2x$ and $g(x) = \sin x$ where $k > 1$.

    The diagram shows the graphs of $y = f(x)$ and $y = g(x)$ intersecting at O, A, B, C and D.

    Show that, at A and C, $\cos x = \dfrac{1}{2k}$.

    5

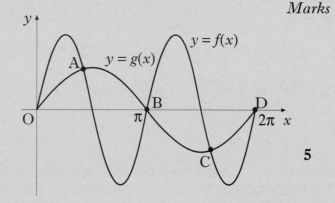

9. The value V (in £ million) of a cruise ship t years after launch is given by the formula $V = 252e^{-0\cdot06335t}$.

    (a) What was its value when launched?

    1

    (b) The owners decide to sell the ship once its value falls below £20 million. After how many years will it be sold?

    4

10. Vectors **a** and **c** are represented by two sides of an equilateral triangle with sides of length 3 units, as shown in the diagram.

    Vector **b** is 2 units long and **b** is perpendicular to both **a** and **c**.

    Evaluate the scalar product **a**.(**a** + **b** + **c**).

    4

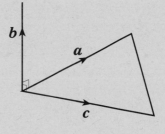

11. (a) Show that $x = -1$ is a solution of the cubic equation $x^3 + px^2 + px + 1 = 0$.

    1

    (b) Hence find the range of values of p for which all the roots of the cubic equation are real.

    7

[END OF QUESTION PAPER]

[BLANK PAGE]

2006 | Higher

[BLANK PAGE]

X100/301

NATIONAL QUALIFICATIONS 2006	FRIDAY, 19 MAY 9.00 AM – 10.10 AM	MATHEMATICS HIGHER Units 1, 2 and 3 Paper 1 (Non-calculator)

Read Carefully

1   **Calculators may <u>NOT</u> be used in this paper.**

2   Full credit will be given only where the solution contains appropriate working.

3   Answers obtained by readings from scale drawings will not receive any credit.

SCOTTISH QUALIFICATIONS AUTHORITY

FORMULAE LIST

Circle:

The equation $x^2 + y^2 + 2gx + 2fy + c = 0$ represents a circle centre $(-g, -f)$ and radius $\sqrt{g^2 + f^2 - c}$.

The equation $(x - a)^2 + (y - b)^2 = r^2$ represents a circle centre (a, b) and radius r.

Scalar Product: $\quad \boldsymbol{a.b} = |\boldsymbol{a}|\,|\boldsymbol{b}|\cos\theta$, where θ is the angle between $\boldsymbol{a}$ and $\boldsymbol{b}$

$\quad$ or $\quad \boldsymbol{a.b} = a_1b_1 + a_2b_2 + a_3b_3$ where $\boldsymbol{a} = \begin{pmatrix} a_1 \\ a_2 \\ a_3 \end{pmatrix}$ and $\boldsymbol{b} = \begin{pmatrix} b_1 \\ b_2 \\ b_3 \end{pmatrix}$.

Trigonometric formulae:

$$\sin(A \pm B) = \sin A \cos B \pm \cos A \sin B$$
$$\cos(A \pm B) = \cos A \cos B \mp \sin A \sin B$$
$$\sin 2A = 2\sin A \cos A$$
$$\cos 2A = \cos^2 A - \sin^2 A$$
$$= 2\cos^2 A - 1$$
$$= 1 - 2\sin^2 A$$

Table of standard derivatives:

$f(x)$	$f'(x)$
$\sin ax$	$a\cos ax$
$\cos ax$	$-a\sin ax$

Table of standard integrals:

$f(x)$	$\int f(x)\,dx$
$\sin ax$	$-\dfrac{1}{a}\cos ax + C$
$\cos ax$	$\dfrac{1}{a}\sin ax + C$

ALL questions should be attempted.

Marks

1. Triangle ABC has vertices A(−1, 12), B(−2, −5) and C(7, −2).

   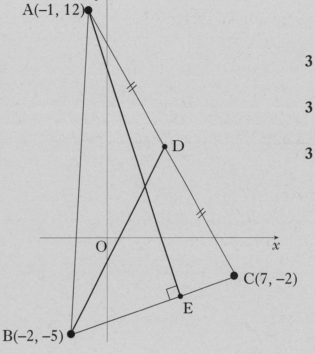

   (a) Find the equation of the median BD.

   3

   (b) Find the equation of the altitude AE.

   3

   (c) Find the coordinates of the point of intersection of BD and AE.

   3

2. A circle has centre C(−2, 3) and passes through P(1, 6).

   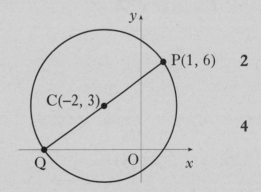

   (a) Find the equation of the circle.

   2

   (b) PQ is a diameter of the circle. Find the equation of the tangent to this circle at Q.

   4

3. Two functions f and g are defined by $f(x) = 2x + 3$ and $g(x) = 2x − 3$, where x is a real number.

   (a) Find expressions for:

   (i) $f(g(x))$;

   (ii) $g(f(x))$.

   3

   (b) Determine the least possible value of the product $f(g(x)) \times g(f(x))$.

   2

[Turn over

Marks

4. A sequence is defined by the recurrence relation $u_{n+1} = 0 \cdot 8u_n + 12$, $u_0 = 4$.

  (*a*) State why this sequence has a limit.   **1**

  (*b*) Find this limit.   **2**

5. A function f is defined by $f(x) = (2x - 1)^5$.

  Find the coordinates of the stationary point on the graph with equation $y = f(x)$
  and determine its nature.   **7**

6. The graph shown has equation $y = x^3 - 6x^2 + 4x + 1$.

  The total shaded area is bounded
  by the curve, the x-axis, the
  y-axis and the line $x = 2$.

  (*a*) Calculate the shaded area
    labelled S.   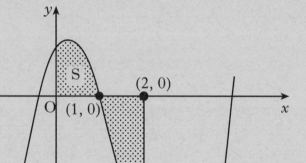   **4**

  (*b*) Hence find the total
    shaded area.   **3**

7. Solve the equation $\sin x° - \sin 2x° = 0$ in the interval $0 \leq x \leq 360$.   **4**

8. (*a*) Express $2x^2 + 4x - 3$ in the form $a(x + b)^2 + c$.   **3**

  (*b*) Write down the coordinates of the turning point on the parabola with
    equation $y = 2x^2 + 4x - 3$.   **1**

Marks

9. u and v are vectors given by $u = \begin{pmatrix} k^3 \\ 1 \\ k+2 \end{pmatrix}$ and $v = \begin{pmatrix} 1 \\ 3k^2 \\ -1 \end{pmatrix}$, where $k > 0$.

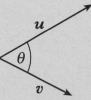

(a) If $u.v = 1$, show that $k^3 + 3k^2 - k - 3 = 0$.   2

(b) Show that $(k + 3)$ is a factor of $k^3 + 3k^2 - k - 3$ and hence factorise $k^3 + 3k^2 - k - 3$ fully.   5

(c) Deduce the only possible value of k.   1

(d) The angle between u and v is θ. Find the exact value of $\cos \theta$.   3

10. Two variables, x and y, are connected by the law $y = a^x$. The graph of $\log_4 y$ against x is a straight line passing through the origin and the point A(6, 3). Find the value of a.   4

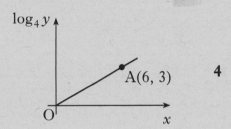

[END OF QUESTION PAPER]

[BLANK PAGE]

X100/303

NATIONAL QUALIFICATIONS 2006	FRIDAY, 19 MAY 10.30 AM – 12.00 NOON	**MATHEMATICS** HIGHER Units 1, 2 and 3 Paper 2

Read Carefully

1 **Calculators may be used in this paper.**

2 Full credit will be given only where the solution contains appropriate working.

3 Answers obtained by readings from scale drawings will not receive any credit.

SCOTTISH QUALIFICATIONS AUTHORITY

FORMULAE LIST

Circle:

The equation $x^2 + y^2 + 2gx + 2fy + c = 0$ represents a circle centre $(-g, -f)$ and radius $\sqrt{g^2 + f^2 - c}$.

The equation $(x - a)^2 + (y - b)^2 = r^2$ represents a circle centre (a, b) and radius r.

Scalar Product: $a.b = |a|\,|b| \cos \theta$, where θ is the angle between a and b

or $a.b = a_1 b_1 + a_2 b_2 + a_3 b_3$ where $a = \begin{pmatrix} a_1 \\ a_2 \\ a_3 \end{pmatrix}$ and $b = \begin{pmatrix} b_1 \\ b_2 \\ b_3 \end{pmatrix}$.

Trigonometric formulae:

$$\sin (A \pm B) = \sin A \cos B \pm \cos A \sin B$$
$$\cos (A \pm B) = \cos A \cos B \mp \sin A \sin B$$
$$\sin 2A = 2\sin A \cos A$$
$$\cos 2A = \cos^2 A - \sin^2 A$$
$$= 2\cos^2 A - 1$$
$$= 1 - 2\sin^2 A$$

Table of standard derivatives:

$f(x)$	$f'(x)$
$\sin ax$	$a \cos ax$
$\cos ax$	$-a \sin ax$

Table of standard integrals:

$f(x)$	$\int f(x)\,dx$
$\sin ax$	$-\dfrac{1}{a} \cos ax + C$
$\cos ax$	$\dfrac{1}{a} \sin ax + C$

ALL questions should be attempted.

Marks

1. PQRS is a parallelogram. P is the point (2, 0), S is (4, 6) and Q lies on the *x*-axis, as shown.

   The diagonal QS is perpendicular to the side PS.

   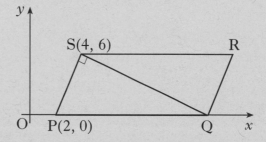

   (*a*) Show that the equation of QS is $x + 3y = 22$.　　4

   (*b*) Hence find the coordinates of Q and R.　　2

2. Find the value of *k* such that the equation $kx^2 + kx + 6 = 0$, $k \neq 0$, has equal roots.　　4

3. The parabola with equation $y = x^2 - 14x + 53$ has a tangent at the point P(8, 5).

   (*a*) Find the equation of this tangent.　　4

   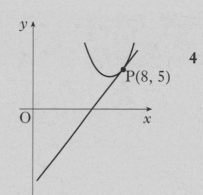

   (*b*) Show that the tangent found in (*a*) is also a tangent to the parabola with equation $y = -x^2 + 10x - 27$ and find the coordinates of the point of contact Q.　　5

   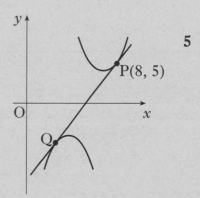

4. The circles with equations $(x - 3)^2 + (y - 4)^2 = 25$ and $x^2 + y^2 - kx - 8y - 2k = 0$ have the same centre.

   Determine the radius of the larger circle.　　5

Marks

5. The curve $y = f(x)$ is such that $\dfrac{dy}{dx} = 4x - 6x^2$. The curve passes through the point $(-1, 9)$. Express y in terms of x.

    4

6. P is the point $(-1, 2, -1)$ and Q is $(3, 2, -4)$.

  (*a*) Write down $\overrightarrow{PQ}$ in component form.

    1

  (*b*) Calculate the length of $\overrightarrow{PQ}$.

    1

  (*c*) Find the components of a unit vector which is parallel to $\overrightarrow{PQ}$.

    1

7. The diagram shows the graph of a function $y = f(x)$.

  Copy the diagram and on it sketch the graphs of:

  (*a*) $y = f(x - 4)$;

    2

  (*b*) $y = 2 + f(x - 4)$.

    2

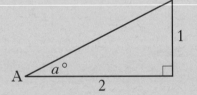

Q(–4, 5) $y = f(x)$ P(1, a)

8. The diagram shows a right-angled triangle with height 1 unit, base 2 units and an angle of $a°$ at A.

  (*a*) Find the exact values of:

    (i) $\sin a°$;

    (ii) $\sin 2a°$.

    4

  (*b*) By expressing $\sin 3a°$ as $\sin(2a + a)°$, find the exact value of $\sin 3a°$.

    4

9. If $y = \dfrac{1}{x^3} - \cos 2x$, $x \neq 0$, find $\dfrac{dy}{dx}$.

    4

10. A curve has equation $y = 7\sin x - 24\cos x$.

  (*a*) Express $7\sin x - 24\cos x$ in the form $k\sin(x - a)$ where $k > 0$ and $0 \leq a \leq \dfrac{\pi}{2}$

    4

  (*b*) Hence find, in the interval $0 \leq x \leq \pi$, the x-coordinate of the point on the curve where the gradient is 1.

    3

Marks

11. It is claimed that a wheel is made from wood which is over 1000 years old.

    To test this claim, carbon dating is used.

    The formula $A(t) = A_0 e^{-0.000124t}$ is used to determine the age of the wood, where A_0 is the amount of carbon in any living tree, $A(t)$ is the amount of carbon in the wood being dated and t is the age of the wood in years.

    For the wheel it was found that $A(t)$ was 88% of the amount of carbon in a living tree.

    Is the claim true?   5

12. PQRS is a rectangle formed according to the following conditions:

    • it is bounded by the lines $x = 6$ and $y = 12$
    • P lies on the curve with equation $y = \dfrac{8}{x}$ between (1, 8) and (4, 2)
    • R is the point (6, 12).

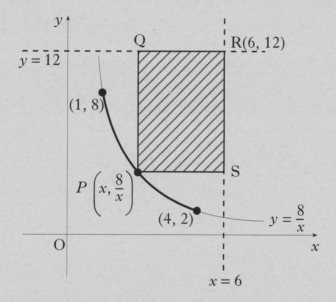

    (a) (i) Express the lengths of PS and RS in terms of x, the x-coordinate of P.

        (ii) Hence show that the area, A square units, of PQRS is given by

        $$A = 80 - 12x - \frac{48}{x}.$$ 3

    (b) Find the greatest and least possible values of A and the corresponding values of x for which they occur.   8

    [END OF QUESTION PAPER]

[BLANK PAGE]

[BLANK PAGE]

X100/301

NATIONAL QUALIFICATIONS 2007	TUESDAY, 15 MAY 9.00 AM – 10.10 AM	**MATHEMATICS** HIGHER Units 1, 2 and 3 Paper 1 (Non-calculator)

Read Carefully

1   **Calculators may <u>NOT</u> be used in this paper.**

2   Full credit will be given only where the solution contains appropriate working.

3   Answers obtained by readings from scale drawings will not receive any credit.

Scottish
Qualifications
Authority

FORMULAE LIST

Circle:

The equation $x^2 + y^2 + 2gx + 2fy + c = 0$ represents a circle centre $(-g, -f)$ and radius $\sqrt{g^2 + f^2 - c}$.

The equation $(x - a)^2 + (y - b)^2 = r^2$ represents a circle centre (a, b) and radius r.

Scalar Product: $\quad a.b = |a|\,|b| \cos \theta$, where θ is the angle between a and b

$$\text{or} \quad a.b = a_1b_1 + a_2b_2 + a_3b_3 \text{ where } a = \begin{pmatrix} a_1 \\ a_2 \\ a_3 \end{pmatrix} \text{ and } b = \begin{pmatrix} b_1 \\ b_2 \\ b_3 \end{pmatrix}.$$

Trigonometric formulae:

$$\sin (A \pm B) = \sin A \cos B \pm \cos A \sin B$$
$$\cos (A \pm B) = \cos A \cos B \mp \sin A \sin B$$
$$\sin 2A = 2\sin A \cos A$$
$$\cos 2A = \cos^2 A - \sin^2 A$$
$$= 2\cos^2 A - 1$$
$$= 1 - 2\sin^2 A$$

Table of standard derivatives:

$f(x)$	$f'(x)$
$\sin ax$	$a \cos ax$
$\cos ax$	$-a \sin ax$

Table of standard integrals:

$f(x)$	$\int f(x)dx$
$\sin ax$	$-\dfrac{1}{a} \cos ax + C$
$\cos ax$	$\dfrac{1}{a} \sin ax + C$

ALL questions should be attempted.

Marks

1. Find the equation of the line through the point (–1, 4) which is parallel to the line with equation $3x - y + 2 = 0$.

3

2. Relative to a suitable coordinate system A and B are the points (–2, 1, –1) and (1, 3, 2) respectively.

   A, B and C are collinear points and C is positioned such that BC = 2AB.

   Find the coordinates of C.

   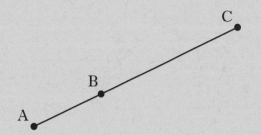

4

3. Functions f and g, defined on suitable domains, are given by $f(x) = x^2 + 1$ and $g(x) = 1 - 2x$.

   Find:

   (a) $g(f(x))$;

2

   (b) $g(g(x))$.

2

4. Find the range of values of k such that the equation $kx^2 - x - 1 = 0$ has no real roots.

4

5. The large circle has equation $x^2 + y^2 - 14x - 16y + 77 = 0$.

   Three congruent circles with centres A, B and C are drawn inside the large circle with the centres lying on a line parallel to the x-axis.

   This pattern is continued, as shown in the diagram.

   Find the equation of the circle with centre D.

   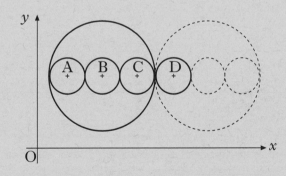

5

[Turn over

Marks

6. Solve the equation $\sin 2x° = 6\cos x°$ for $0 \leq x \leq 360$.   **4**

7. A sequence is defined by the recurrence relation

$$u_{n+1} = \tfrac{1}{4}u_n + 16,\ u_0 = 0.$$

(a) Calculate the values of u_1, u_2 and u_3.   **3**

Four terms of this sequence, u_1, u_2, u_3 and u_4 are plotted as shown in the graph.

As $n \to \infty$, the points on the graph approach the line $u_n = k$, where k is the limit of this sequence.

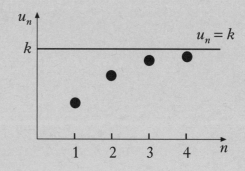

(b) (i) Give a reason why this sequence has a limit.

(ii) Find the exact value of k.   **3**

8. The diagram shows a sketch of the graph of $y = x^3 - 4x^2 + x + 6$.

(a) Show that the graph cuts the x-axis at $(3, 0)$.   **1**

(b) Hence or otherwise find the coordinates of A.   **3**

(c) Find the shaded area.   **5**

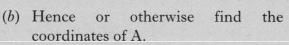

9. A function f is defined by the formula $f(x) = 3x - x^3$.

(a) Find the exact values where the graph of $y = f(x)$ meets the x- and y-axes.   **2**

(b) Find the coordinates of the stationary points of the function and determine their nature.   **7**

(c) Sketch the graph of $y = f(x)$.   **1**

Marks

10. Given that $y = \sqrt{3x^2 + 2}$, find $\frac{dy}{dx}$. **3**

11. (a) Express $f(x) = \sqrt{3}\cos x + \sin x$ in the form $k\cos(x - a)$, where $k > 0$ and $0 < a < \frac{\pi}{2}$. **4**

    (b) Hence or otherwise sketch the graph of $y = f(x)$ in the interval $0 \le x \le 2\pi$. **4**

[END OF QUESTION PAPER]

[BLANK PAGE]

X100/303

NATIONAL
QUALIFICATIONS
2007

TUESDAY, 15 MAY
10.30 AM – 12.00 NOON

MATHEMATICS
HIGHER
Units 1, 2 and 3
Paper 2

Read Carefully

1 **Calculators may be used in this paper.**

2 Full credit will be given only where the solution contains appropriate working.

3 Answers obtained by readings from scale drawings will not receive any credit.

SCOTTISH
QUALIFICATIONS
AUTHORITY

LI X100/303 6/27370

FORMULAE LIST

Circle:

The equation $x^2 + y^2 + 2gx + 2fy + c = 0$ represents a circle centre $(-g, -f)$ and radius $\sqrt{g^2 + f^2 - c}$.

The equation $(x - a)^2 + (y - b)^2 = r^2$ represents a circle centre (a, b) and radius r.

Scalar Product: $a.b = |a|\,|b|\cos\theta$, where θ is the angle between a and b

or $a.b = a_1b_1 + a_2b_2 + a_3b_3$ where $a = \begin{pmatrix} a_1 \\ a_2 \\ a_3 \end{pmatrix}$ and $b = \begin{pmatrix} b_1 \\ b_2 \\ b_3 \end{pmatrix}$.

Trigonometric formulae: $\sin(A \pm B) = \sin A \cos B \pm \cos A \sin B$

$\cos(A \pm B) = \cos A \cos B \mp \sin A \sin B$

$\sin 2A = 2\sin A \cos A$

$\cos 2A = \cos^2 A - \sin^2 A$

$= 2\cos^2 A - 1$

$= 1 - 2\sin^2 A$

Table of standard derivatives:

$f(x)$	$f'(x)$
$\sin ax$	$a\cos ax$
$\cos ax$	$-a\sin ax$

Table of standard integrals:

$f(x)$	$\int f(x)\,dx$
$\sin ax$	$-\dfrac{1}{a}\cos ax + C$
$\cos ax$	$\dfrac{1}{a}\sin ax + C$

ALL questions should be attempted.

Marks

1. OABCDEFG is a cube with side 2 units, as shown in the diagram.

   B has coordinates (2, 2, 0).

   P is the centre of face OCGD and Q is the centre of face CBFG.

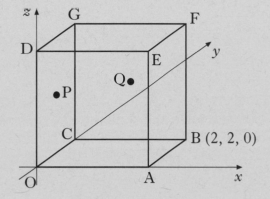

   (*a*) Write down the coordinates of G.   1

   (*b*) Find p and q, the position vectors of points P and Q.   2

   (*c*) Find the size of angle POQ.   5

2. The diagram shows two right-angled triangles with angles *c* and *d* marked as shown.

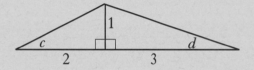

   (*a*) Find the exact value of $\sin(c + d)$.   4

   (*b*) (i) Find the exact value of $\sin 2c$.

   (ii) Show that $\cos 2d$ has the same exact value.   4

3. Show that the line with equation $y = 6 - 2x$ is a tangent to the circle with equation $x^2 + y^2 + 6x - 4y - 7 = 0$ and find the coordinates of the point of contact of the tangent and the circle.   6

4. The diagram shows part of the graph of a function whose equation is of the form $y = a\sin(bx^\circ) + c$.

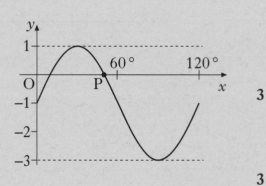

   (*a*) Write down the values of *a*, *b* and *c*.   3

   (*b*) Determine the exact value of the *x*-coordinate of P, the point where the graph intersects the *x*-axis as shown in the diagram.   3

Marks

5. A circle centre C is situated so that it touches the parabola with equation $y = \frac{1}{2}x^2 - 8x + 34$ at P and Q.

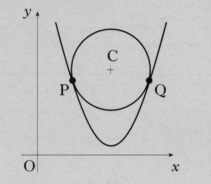

   (a) The gradient of the tangent to the parabola at Q is 4. Find the coordinates of Q.

5

   (b) Find the coordinates of P.

2

   (c) Find the coordinates of C, the centre of the circle.

2

6. A householder has a garden in the shape of a right-angled isosceles triangle.

It is intended to put down a section of rectangular wooden decking at the side of the house, as shown in the diagram.

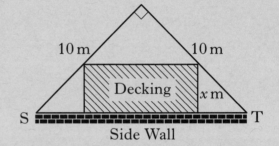

   (a)   (i) Find the exact value of ST.

       (ii) Given that the breadth of the decking is x metres, show that the area of the decking, A square metres, is given by

$$A = \left(10\sqrt{2}\right)x - 2x^2.$$

3

   (b) Find the dimensions of the decking which maximises its area.

5

7. Find the value of $\int_0^2 \sin(4x + 1)dx$.

4

8. The curve with equation $y = \log_3(x - 1) - 2 \cdot 2$, where $x > 1$, cuts the x-axis at the point $(a, 0)$.

Find the value of a.

4

Marks

9. The diagram shows the graph of $y = a^x$, $a > 1$.

   On separate diagrams, sketch the graphs of:

   (a) $y = a^{-x}$;   2

   (b) $y = a^{1-x}$.   2

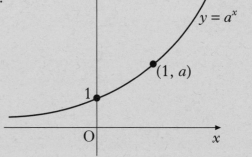

10. The diagram shows the graphs of a cubic function $y = f(x)$ and its derived function $y = f'(x)$.

    Both graphs pass through the point $(0, 6)$.

    The graph of $y = f'(x)$ also passes through the points $(2, 0)$ and $(4, 0)$.

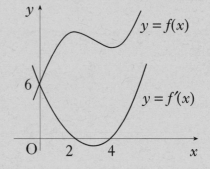

    (a) Given that $f'(x)$ is of the form $k(x - a)(x - b)$:

        (i) write down the values of a and b;

        (ii) find the value of k.   3

    (b) Find the equation of the graph of the cubic function $y = f(x)$.   4

11. Two variables x and y satisfy the equation $y = 3 \times 4^x$.

    (a) Find the value of a if $(a, 6)$ lies on the graph with equation $y = 3 \times 4^x$.   1

    (b) If $(-\frac{1}{2}, b)$ also lies on the graph, find b.   1

    (c) A graph is drawn of $\log_{10} y$ against x. Show that its equation will be of the form $\log_{10} y = Px + Q$ and state the gradient of this line.   4

[END OF QUESTION PAPER]

[BLANK PAGE]

2007 | SQP

[BLANK PAGE]

[C100/SQP321]

Mathematics
Higher
Paper 1
Specimen Question Paper
(for examinations from Diet 2008 onwards)

NATIONAL
QUALIFICATIONS

Read carefully

Calculators may <u>NOT</u> be used in this paper.

Section A – Questions 1–20 (40 marks)

Instructions for completion of **Section A** are given on page two.

For this section of the examination you must use an **HB pencil**.

Section B (30 marks)

1   Full credit will be given only where the solution contains appropriate working.

2   Answers obtained by readings from scale drawings will not receive any credit.

Read carefully

1 Check that the answer sheet provided is for **Mathematics Higher (Section A)**.

2 For this section of the examination you must use an **HB pencil** and, where necessary, an eraser.

3 Check that the answer sheet you have been given has **your name**, **date of birth**, **SCN** (Scottish Candidate Number) and **Centre Name** printed on it.

   Do not change any of these details.

4 If any of this information is wrong, tell the Invigilator immediately.

5 If this information is correct, **print** your name and seat number in the boxes provided.

6 The answer to each question is **either** A, B, C or D. Decide what your answer is, then, using your pencil, put a horizontal line in the space provided (see sample question below).

7 There is **only one correct** answer to each question.

8 Rough working should **not** be done on your answer sheet.

9 At the end of the exam, put the **answer sheet for Section A inside the front cover of your answer book**.

Sample Question

A curve has equation $y = x^3 - 4x$.

What is the gradient at the point where $x = 2$?

        A 8

        B 1

        C 0

        D –4

The correct answer is **A**—8. The answer **A** has been clearly marked in **pencil** with a horizontal line (see below).

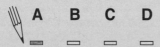

Changing an answer

If you decide to change your answer, carefully erase your first answer and using your pencil, fill in the answer you want. The answer below has been changed to **D**.

        A   B   C   D

FORMULAE LIST

Circle:

The equation $x^2 + y^2 + 2gx + 2fy + c = 0$ represents a circle centre $(-g, -f)$ and radius $\sqrt{g^2 + f^2 - c}$.

The equation $(x - a)^2 + (y - b)^2 = r^2$ represents a circle centre (a, b) and radius r.

Scalar Product: $\quad$ $\boldsymbol{a}.\boldsymbol{b} = |\boldsymbol{a}|\,|\boldsymbol{b}| \cos \theta$, where θ is the angle between $\boldsymbol{a}$ and $\boldsymbol{b}$

or $\quad$ $\boldsymbol{a}.\boldsymbol{b} = a_1 b_1 + a_2 b_2 + a_3 b_3$ where $\boldsymbol{a} = \begin{pmatrix} a_1 \\ a_2 \\ a_3 \end{pmatrix}$ and $\boldsymbol{b} = \begin{pmatrix} b_1 \\ b_2 \\ b_3 \end{pmatrix}$.

Trigonometric formulae:

$$\sin (A \pm B) = \sin A \cos B \pm \cos A \sin B$$
$$\cos (A \pm B) = \cos A \cos B \mp \sin A \sin B$$
$$\sin 2A = 2\sin A \cos A$$
$$\cos 2A = \cos^2 A - \sin^2 A$$
$$= 2\cos^2 A - 1$$
$$= 1 - 2\sin^2 A$$

Table of standard derivatives:

$f(x)$	$f'(x)$
$\sin ax$	$a \cos ax$
$\cos ax$	$-a \sin ax$

Table of standard integrals:

$f(x)$	$\int f(x)\,dx$
$\sin ax$	$-\dfrac{1}{a}\cos ax + C$
$\cos ax$	$\dfrac{1}{a}\sin ax + C$

SECTION A

ALL questions should be attempted.

1. If $y = \dfrac{x^3 - x}{x^2}$, what is $\dfrac{dy}{dx}$?

   A $\dfrac{3x^2 - 1}{2x}$

   B $1 + \dfrac{1}{x^2}$

   C $\dfrac{3}{2}x - \dfrac{1}{2}$

   D $x^3 - x - x^{-2}$

2. Functions f and g are given by $f(x) = 2x - 3$ and $g(x) = x^2$.

   Find an expression for $g(f(x))$.

   A $g(f(x)) = 4x^2 - 12x + 9$
   B $g(f(x)) = x^2 + 2x - 3$
   C $g(f(x)) = 4x - 9$
   D $g(f(x)) = 2x^3 - 3x^2$

3. Find $\displaystyle\int \dfrac{1}{\sqrt[3]{x}}\, dx$.

   A $-\dfrac{3}{2}x^{-\frac{1}{2}} + c$

   B $x^{-3} + c$

   C $\dfrac{3}{2}x^{\frac{2}{3}} + c$

   D $-2x^{-2} + c$

4. A and B have coordinates $(2, 3, -2)$ and $(-1, -4, 0)$ respectively.

What is the distance between A and B?

A $\sqrt{6}$

B $\sqrt{17}$

C $\sqrt{62}$

D $\sqrt{148}$

5. A sequence is defined by the recurrence relation

$$u_{n+1} = 3u_n - 4, \; u_0 = -1.$$

What is the value of u_2?

A -25

B -10

C -4

D -1

6. The diagram shows a sketch of $y = f(x)$.

Which of the diagrams below shows a sketch of $y = -3 - f(x)$?

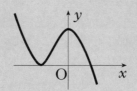

A

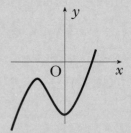

B

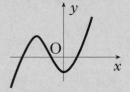

C

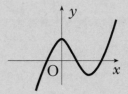

D

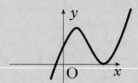

7. Which of the following describes the stationary point on the curve with equation $y = 3(x - 4)^2 - 5$?

A minimum at (4, 5)

B maximum at (4, 5)

C minimum at (4, –5)

D maximum at (4, –5)

8. The diagram shows a right-angled triangle with sides of 1, $2\sqrt{2}$ and 3.

   What is the value of $\sin 2x°$?

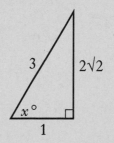

   A $\dfrac{4\sqrt{2}}{9}$

   B $\sqrt{\dfrac{2}{3}}$

   C $\dfrac{4\sqrt{2}}{3}$

   D $\dfrac{3}{\sqrt{2}}$

9. a and b are angles as shown in the diagram.

   What is the value of $\sin(a - b)$?

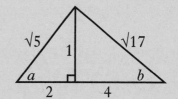

   A $-\dfrac{7}{\sqrt{85}}$

   B $\dfrac{2}{\sqrt{85}}$

   C $\dfrac{1}{\sqrt{5}} + \dfrac{1}{\sqrt{17}}$

   D $\dfrac{1}{\sqrt{5}} - \dfrac{1}{\sqrt{17}}$

10. A circle has equation $x^2 + y^2 + 8x - 6y - 12 = 0$.

What is the radius of this circle?

A $\sqrt{2}$

B $\sqrt{19}$

C $\sqrt{37}$

D $\sqrt{88}$

11. The points P(1, 3, 7), Q(5, 13, 13) and R(s, 33, 25) are collinear as shown in the diagram.

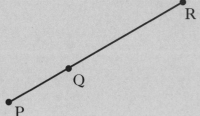

What is the value of s?

A   9

B   10

C   13

D   31

12. If $2x^2 - 12x + 11$ is expressed in the form $2(x - b)^2 + c$, what is the value of c?

A   −25

B   −7

C   11

D   23

13. The curve $y = f(x)$ is such that $\dfrac{dy}{dx} = 3x^2 + 9x + 1$ and the curve passes through the origin.

What is the equation of the curve?

A $y = x^3 + \dfrac{9}{2}x^2 + x$

B $y = 6x^3 + 9x^2$

C $y = 3x^3 + 9x^2 + x + 1$

D $y = 6x + 9$

14. For what value of k does the equation $x^2 - 3x + k = 0$ have equal roots?

     A $-\dfrac{9}{4}$

     B $-\dfrac{1}{12}$

     C 0

     D $\dfrac{9}{4}$

15. The point P(-1, 2) lies on the circle with equation $x^2 + y^2 - 6x - 8y + 5 = 0$.

What is the gradient of the tangent at P?

     A -2

     B $-\dfrac{1}{3}$

     C $\dfrac{6}{7}$

     D $\dfrac{1}{2}$

16. What is the value of $\displaystyle\int_0^{\frac{\pi}{6}} 4\cos 2x\, dx$?

     A -2

     B $-\sqrt{\dfrac{3}{2}}$

     C $\sqrt{3}$

     D 4

17. The graph shown in the diagram has equation of the form $y = \sin(px) + q$.

What are the values of p and q?

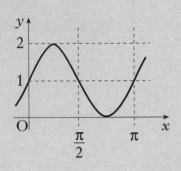

	p	q
A	2	1
B	$\frac{1}{2}$	1
C	2	2
D	$\frac{1}{2}$	2

18. The vectors $\boldsymbol{a}$, $\boldsymbol{b}$ and $\boldsymbol{c}$ are represented by the sides of a right-angled triangle as shown in the diagram.

$|\boldsymbol{a}| = 3$ and $|\boldsymbol{c}| = 5$.

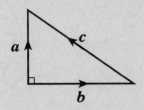

Here are two statements about these vectors:

(1) $\boldsymbol{a.c} = 9$

(2) $\boldsymbol{a.b} = -1$

Which of the following is true?

A   neither statement is correct

B   only statement (1) is correct

C   only statement (2) is correct

D   both statements are correct

19. If $\log_3 t = 2 + \log_3 5$, what is the value of t?

A   7

B   10

C   25

D   45

20. If $3^k = e^4$, find an expression for k.

A $k = \sqrt[3]{4^e}$

B $k = \dfrac{e^4}{3}$

C $k = 4 \,/\, \log_e 3$

D $k = 1 \,/\, \log_e 3$

[END OF SECTION A]

SECTION B

ALL questions should be attempted.

Marks

21. A firm cleans the factory floor on a daily basis with disinfectant. It has a choice of two products, either "A" or "B".

    Product A removes 70% of all germs but during the next 24 hours, 300 "new" germs per sq unit are estimated to appear.

    Product B removes 80% of all germs but during the next 24 hours, 350 "new" germs per sq unit are estimated to appear.

    For product A, let u_n represent the number of germs per sq unit on the floor immediately before disinfecting for the nth time.

    For product B, let v_n represent the number of germs per sq unit on the floor immediately before disinfecting for the nth time.

    (a) Write down a recurrence relation for each product to show the number of germs per sq unit present prior to disinfecting.   2

    (b) Determine which product is more effective in the long term.   4

22. (a) Find the stationary points on the curve with equation $y = x^3 - 9x^2 + 24x - 20$ and justify their nature.   7

    (b) (i) Show that $(x - 2)^2(x - 5) = x^3 - 9x^2 + 24x - 20$.

    (ii) Hence sketch the graph of $y = x^3 - 9x^2 + 24x - 20$.   4

23. The diagram shows a sketch of functions f and g where $f(x) = x^3 + 5x^2 - 36x + 32$ and $g(x) = -x^2 + x + 2$.

    The two graphs intersect at the points A, B and C.

    Determine the x-coordinate of each of these three points.   8

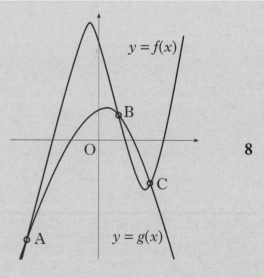

24. Find the solution(s) of the equation $\sin^2 p - \sin p + 1 = \cos^2 p$ for $\frac{\pi}{2} < p < \pi$.   5

[END OF SECTION B]

[END OF QUESTION PAPER]

[C100/SQP321]

Mathematics
Higher
Paper 2
Specimen Question Paper
(for examinations from Diet 2008 onwards)

NATIONAL
QUALIFICATIONS

Read Carefully

1 **Calculators may be used in this paper.**

2 Full credit will be given only where the solution contains appropriate working.

3 Answers obtained by readings from scale drawings will not receive any credit.

SCOTTISH
QUALIFICATIONS
AUTHORITY
©

FORMULAE LIST

Circle:

The equation $x^2 + y^2 + 2gx + 2fy + c = 0$ represents a circle centre $(-g, -f)$ and radius $\sqrt{g^2 + f^2 - c}$.

The equation $(x - a)^2 + (y - b)^2 = r^2$ represents a circle centre (a, b) and radius r.

Scalar Product: $a.b = |a|\,|b| \cos \theta$, where θ is the angle between a and b

or $a.b = a_1b_1 + a_2b_2 + a_3b_3$ where $a = \begin{pmatrix} a_1 \\ a_2 \\ a_3 \end{pmatrix}$ and $b = \begin{pmatrix} b_1 \\ b_2 \\ b_3 \end{pmatrix}$.

Trigonometric formulae:

$$\sin (A \pm B) = \sin A \cos B \pm \cos A \sin B$$
$$\cos (A \pm B) = \cos A \cos B \mp \sin A \sin B$$
$$\sin 2A = 2\sin A \cos A$$
$$\cos 2A = \cos^2 A - \sin^2 A$$
$$= 2\cos^2 A - 1$$
$$= 1 - 2\sin^2 A$$

Table of standard derivatives:

$f(x)$	$f'(x)$
$\sin ax$	$a \cos ax$
$\cos ax$	$-a \sin ax$

Table of standard integrals:

$f(x)$	$\int f(x)\,dx$
$\sin ax$	$-\dfrac{1}{a} \cos ax + C$
$\cos ax$	$\dfrac{1}{a} \sin ax + C$

ALL questions should be attempted.

Marks

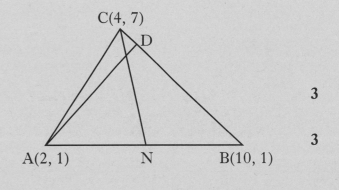

1. Triangle ABC has coordinates A(2, 1), B(10, 1) and C(4, 7).

   (a) Find the equation of the median CN.

   3

   (b) Find the equation of the altitude AD.

   3

   (c) The median from (a) and the altitude from (b) intersect at P. Find the coordinates of P.

   3

   (d) The point Q lies on AB and has coordinates (8, 1).

   Show that PQ is parallel to BC.

   2

2. The diagram shows a wire framework in the shape of a cuboid with the edges parallel to the axes.

   Relative to these axes, A, B, C and H have coordinates (1, 3, 4), (2, 3, 4), (2, 7, 4) and (1, 7, 9) respectively.

   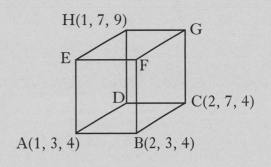

   (a) State the lengths of AB, AD and AE.

   1

   (b) Write down the components of $\overrightarrow{HB}$ and $\overrightarrow{HC}$ and hence or otherwise calculate the size of angle BHC.

   7

3. (a) Express $5\sin x° - 12\cos x°$ in the form $k\sin(x - a)°$ where $k > 0$ and $0 < a < 360$.

   4

   (b) Hence solve the equation $5\sin x° - 12\cos x° = 6·5$ in the interval $0 < x < 360$.

   3

Marks

4. The diagram shows a parabola with equation $y = 2x^2 - 2x + 3$.

   A tangent to the parabola has been drawn at P(1, 3).

   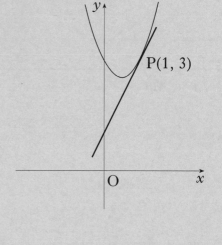

   (a) Find the equation of this tangent. **4**

   A circle has equation $x^2 + y^2 + 8y + 11 = 0$.

   (b) Show that the line from (a) is also a tangent to this circle and state the coordinates of the point of contact Q. **6**

   (c) Determine the ratio in which the y-axis cuts the line QP. **3**

5. The diagram shows a curve with equation $y = x^2$ and a straight line with equation $y = 6x + 16$ intersecting the curve at P and Q.

   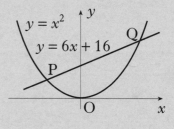

   (a) Calculate the exact value of the area enclosed by the curve and the straight line. **7**

   The second diagram shows a third point, R, lying on the curve between P and Q.

   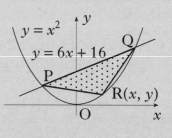

   (b) The area, A, of triangle PQR, is given by $A(x) = -5x^2 + 30x + 80$.

   Determine the maximum area of this triangle, and express your answer as a fraction of the area enclosed by the curve and the straight line. **4**

6. Radium decays exponentially and its half-life is 1600 years.

   If A_0 represents the amount of radium in a sample to start with and $A(t)$ represents the amount remaining after t years, then $A(t) = A_0 e^{-kt}$.

   (a) Determine the value of k, correct to 4 significant figures. **3**

   (b) Hence find what percentage, to the nearest whole number, of the original amount of radium will be remaining after 3200 years. **2**

Marks

7. Triangle ABC is right-angled at A and BD is the bisector of angle ABC.

   AB = 6 units and CB = 10 units.

   Determine the exact value of BD, expressing your answer in its simplest form.

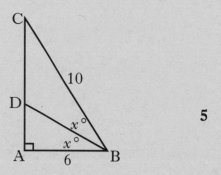

5

[*END OF QUESTION PAPER*]

[BLANK PAGE]

Official SQA Past Papers: Higher Maths 2007 SQP

[BLANK PAGE]

[C100/SQP328]

Mathematics
Higher
Paper 1
Specimen Question Paper
Example 2 based on 2004 Examination Paper
(for examinations from Diet 2008 onwards)

NATIONAL
QUALIFICATIONS

Read carefully

Calculators may <u>NOT</u> be used in this paper.

Section A – Questions 1–20 (40 marks)

Instructions for completion of **Section A** are given on page two.

For this section of the examination you must use an **HB pencil**.

Section B (30 marks)

1   Full credit will be given only where the solution contains appropriate working.

2   Answers obtained by readings from scale drawings will not receive any credit.

SCOTTISH
QUALIFICATIONS
AUTHORITY
©

Read carefully

1 Check that the answer sheet provided is for **Mathematics Higher (Section A)**.

2 For this section of the examination you must use an **HB pencil** and, where necessary, an eraser.

3 Check that the answer sheet you have been given has **your name**, **date of birth**, **SCN** (Scottish Candidate Number) and **Centre Name** printed on it.

   Do not change any of these details.

4 If any of this information is wrong, tell the Invigilator immediately.

5 If this information is correct, **print** your name and seat number in the boxes provided.

6 The answer to each question is **either** A, B, C or D. Decide what your answer is, then, using your pencil, put a horizontal line in the space provided (see sample question below).

7 There is **only one correct** answer to each question.

8 Rough working should **not** be done on your answer sheet.

9 At the end of the exam, put the **answer sheet for Section A inside the front cover of your answer book**.

Sample Question

A curve has equation $y = x^3 - 4x$.

What is the gradient at the point where $x = 2$?

   A   8

   B   1

   C   0

   D −4

The correct answer is **A**—8. The answer **A** has been clearly marked in **pencil** with a horizontal line (see below).

A   B   C   D

Changing an answer

If you decide to change your answer, carefully erase your first answer and using your pencil, fill in the answer you want. The answer below has been changed to **D**.

A   B   C   D

FORMULAE LIST

Circle:

The equation $x^2 + y^2 + 2gx + 2fy + c = 0$ represents a circle centre $(-g, -f)$ and radius $\sqrt{g^2 + f^2 - c}$.

The equation $(x - a)^2 + (y - b)^2 = r^2$ represents a circle centre (a, b) and radius r.

Scalar Product: $\boldsymbol{a}.\boldsymbol{b} = |\boldsymbol{a}|\,|\boldsymbol{b}| \cos \theta$, where θ is the angle between $\boldsymbol{a}$ and $\boldsymbol{b}$

or $\boldsymbol{a}.\boldsymbol{b} = a_1b_1 + a_2b_2 + a_3b_3$ where $\boldsymbol{a} = \begin{pmatrix} a_1 \\ a_2 \\ a_3 \end{pmatrix}$ and $\boldsymbol{b} = \begin{pmatrix} b_1 \\ b_2 \\ b_3 \end{pmatrix}$.

Trigonometric formulae:
$$\sin (A \pm B) = \sin A \cos B \pm \cos A \sin B$$
$$\cos (A \pm B) = \cos A \cos B \mp \sin A \sin B$$
$$\sin 2A = 2\sin A \cos A$$
$$\cos 2A = \cos^2 A - \sin^2 A$$
$$= 2\cos^2 A - 1$$
$$= 1 - 2\sin^2 A$$

Table of standard derivatives:

$f(x)$	$f'(x)$
$\sin ax$	$a \cos ax$
$\cos ax$	$-a \sin ax$

Table of standard integrals:

$f(x)$	$\int f(x)\,dx$
$\sin ax$	$-\dfrac{1}{a} \cos ax + C$
$\cos ax$	$\dfrac{1}{a} \sin ax + C$

SECTION A

ALL questions should be attempted.

1. The line through P(7, p) and Q(4, −5) has a gradient of 3.

   What is the value of p?

   A   −14

   B   4

   C   6

   D   8

2. A sequence is defined by the recurrence relation $u_{n+1} = u_n + 5$, $u_0 = -3$.

   What is the value of u_2?

   A   3
   B   5
   C   7
   D   9

3. What is the gradient of the line perpendicular to the line with equation $3y = -2x + 1$?

   A   −3

   B   1

   C $\dfrac{3}{2}$

   D   5

4. $f(x) = x^3 - x^2 - 5x - 3$.

   What is the remainder when $f(x)$ is divided by $(x + 3)$?

   A   −24
   B   −3
   C   36
   D   48

5. If $x^2 - 16x + 27$ is written in the form $(x + p)^2 + q$, find the value of q.

   A   −37

   B   11

   C   27

   D   43

6. What is the derivative of $(8 - 2x^2)^{\frac{2}{3}}$?

   A $-\dfrac{8}{3}x(8 - 2x^2)^{-\frac{1}{3}}$

   B $(8 - 4x)^{\frac{2}{3}}$

   C $\dfrac{2}{3}(8 - 4x)^{-\frac{1}{3}}$

   D $\dfrac{3}{5}(8 - 2x^2)^{\frac{5}{3}}$

7. On dividing $f(x)$ by $(x - 1)$, the remainder is zero and the quotient is $x^2 - 4x - 5$. Find $f(x)$ in its fully factorised form.

   A $(x - 1)(x - 1)(x + 5)$

   B $(x + 1)(x - 5)$

   C $(x - 1)(x - 1)$

   D $(x - 1)(x + 1)(x - 5)$

8. A sequence is generated by the recurrence relation $u_{n+1} = 0.4u_n + 3$. What is the limit of this sequence as $n \to \infty$?

   A $\dfrac{1}{5}$

   B $\dfrac{15}{7}$

   C   5

   D $\dfrac{15}{2}$

9. Find all the values of x in the interval $0 < x < 2\pi$ for which $\tan x = -\sqrt{3}$.

A $\dfrac{5\pi}{6}, \dfrac{11\pi}{6}$

B $\dfrac{2\pi}{3}, \dfrac{4\pi}{3}$

C $\dfrac{2\pi}{3}, \dfrac{5\pi}{3}$

D $\dfrac{5\pi}{3}, \dfrac{7\pi}{3}$

10. $P = (-3, 4, 7)$, $Q = (-1, 8, 3)$ and $R = (0, 10, 1)$.

Find the ratio in which Q divides PR.

A $2 : 1$

B $3 : -1$

C $1 : 2$

D $3 : 1$

11. The diagram shows the line OP with equation $2y = x$.

The angle between OP and the positive direction of the x-axis is p°.

Find an expression for angle p.

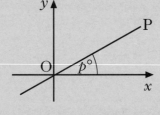

A $\tan^{-1}\tfrac{1}{2}$

B $\tan^{-1}1$

C $\tan^{-1}2$

D $-\tan^{-1}\tfrac{1}{2}$

12. Which one of the following is true for the function g where $g'(x) = x^2 + 2x + 1$?

    A g is never increasing.

    B g is decreasing then increasing.

    C g is increasing then decreasing.

    D g is never decreasing.

13. Simplify $\log_2(x+1) - 2\log_2 3$.

    A $\log_2\left(\dfrac{x+1}{9}\right)$

    B $\log_2(x-8)$

    C $\log_2(x-2)$

    D $\log_2 6(x+1)$

14. The diagram shows the graph of $y = g(x)$.

Which diagram below shows the graph of $y = 3 - g(x)$?

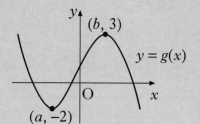

A

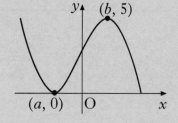

B

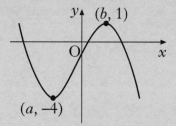

C

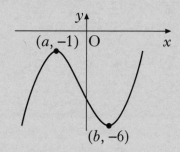

D

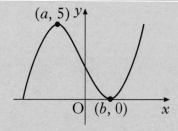

Page eight

15. Points P and Q have coordinates $(1, 3, -1)$ and $(2, 5, 1)$ and T is the midpoint of PQ.

What is the position vector of T?

A $\begin{pmatrix} -\dfrac{3}{2} \\ -4 \\ 0 \end{pmatrix}$

B $\begin{pmatrix} \dfrac{3}{2} \\ 4 \\ 0 \end{pmatrix}$

C $\begin{pmatrix} -\dfrac{1}{2} \\ -1 \\ -1 \end{pmatrix}$

D $\begin{pmatrix} -1 \\ -2 \\ -2 \end{pmatrix}$

16. $A = (-3, 4, 7)$ and $B = (-1, 8, 3)$.

If $\overrightarrow{AD} = 4\overrightarrow{AB}$, what are the coordinates of D?

A $(-9, -8, -13)$

B $(5, -4, 1$

C $(-6, 8, 14)$

D $(5, 20, -9)$

17. The equation of the parabola shown is of the form $y = kx(x - 6)$.

What is the value of k?

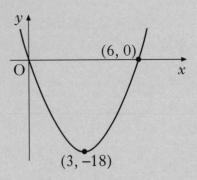

A 0

B $\dfrac{1}{144}$

C 2

D 6

18. Given that $y = 3\cos 5x$, find $\dfrac{dy}{dx}$.

A $15\cos 5x$

B $-15\sin 5x$

C $-15\cos x$

D $3\cos 5$

19. Find $\displaystyle\int (4x + 1)^{\frac{1}{2}}\ dx$.

A $\dfrac{1}{6}(4x + 1)^{\frac{3}{2}} + c$

B $\dfrac{1}{4}(4x + 1) + c$

C $\dfrac{1}{4}(4x + 1)^{\frac{3}{2}} + c$

D $2(4x + 1)^{-\frac{3}{2}} + c$

20. Given that $\int (3x+1)^{-\frac{1}{2}} \, dx = \frac{2}{3}(3x+1)^{\frac{1}{2}} + c$, find $\int_0^1 (3x+1)^{-\frac{1}{2}} \, dx$.

A $\frac{2}{3}$

B $\frac{4}{3}$

C   2

D $\sqrt{2}$

[END OF SECTION A]

SECTION B

ALL questions should be attempted.

Marks

21. (*a*) Find the stationary points on the curve with equation $y = x^3 + 3x^2 - 9x + 5$ and justify their nature.

7

   (*b*) The curve passes through the point $(-5, 0)$. Sketch the curve.

2

22. Solve the equation $\log_x 8 + \log_x 4 = 5$.

4

23. Solve the equation $\sin 2x - \cos x = 0$ for $0 \le x \le 2\pi$.

5

24. In the diagram,

angle DEC = angle CEB = $x°$ and
angle CDE = angle BEA = $90°$.
CD = 1 unit; DE = 3 units.

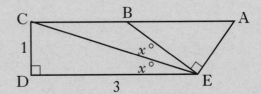

By writing angle DEA in terms of $x°$,
find the exact value of $\cos(\text{D}\hat{\text{E}}\text{A})$.

7

25. The diagram shows a parabola with equation

$$y = 6x(x - 2).$$

This parabola is the graph of $y = f'(x)$.

Given that $f(1) = 4$, find the formula for $f(x)$.

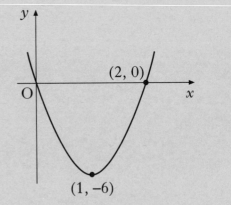

5

[*END OF SECTION B*]

[*END OF QUESTION PAPER*]

[C100/SQP328]

Mathematics
Higher
Paper 2
Specimen Question Paper
Example 2 based on 2004 Examination Paper
(for examinations from Diet 2008 onwards)

NATIONAL
QUALIFICATIONS

Read Carefully

1 **Calculators may be used in this paper.**

2 Full credit will be given only where the solution contains appropriate working.

3 Answers obtained by readings from scale drawings will not receive any credit.

FORMULAE LIST

Circle:

The equation $x^2 + y^2 + 2gx + 2fy + c = 0$ represents a circle centre $(-g, -f)$ and radius $\sqrt{g^2 + f^2 - c}$.

The equation $(x - a)^2 + (y - b)^2 = r^2$ represents a circle centre (a, b) and radius r.

Scalar Product: $\quad a.b = |a|\,|b|\cos\theta$, where θ is the angle between a and b

$$\text{or} \quad a.b = a_1b_1 + a_2b_2 + a_3b_3 \text{ where } a = \begin{pmatrix} a_1 \\ a_2 \\ a_3 \end{pmatrix} \text{ and } b = \begin{pmatrix} b_1 \\ b_2 \\ b_3 \end{pmatrix}.$$

Trigonometric formulae:

$$\sin(A \pm B) = \sin A \cos B \pm \cos A \sin B$$
$$\cos(A \pm B) = \cos A \cos B \mp \sin A \sin B$$
$$\sin 2A = 2\sin A \cos A$$
$$\cos 2A = \cos^2 A - \sin^2 A$$
$$= 2\cos^2 A - 1$$
$$= 1 - 2\sin^2 A$$

Table of standard derivatives:

$f(x)$	$f'(x)$
$\sin ax$	$a\cos ax$
$\cos ax$	$-a\sin ax$

Table of standard integrals:

$f(x)$	$\int f(x)\,dx$
$\sin ax$	$-\dfrac{1}{a}\cos ax + C$
$\cos ax$	$\dfrac{1}{a}\sin ax + C$

ALL questions should be attempted.

Marks

1. Given that $\overrightarrow{QP} = \begin{pmatrix} -1 \\ 3 \\ -2 \end{pmatrix}$ and $\overrightarrow{QR} = \begin{pmatrix} -5 \\ 1 \\ 1 \end{pmatrix}$, find the size of angle PQR.

   5

2. Prove that the roots of the equation $2x^2 + px - 3 = 0$ are real for all values of p.

   4

3. The point P(x, y) lies on the curve with equation $y = 6x^2 - x^3$.

   (a) Find the value of x for which the gradient of the tangent at P is 12.

   5

   (b) Hence find the equation of the tangent at P.

   2

4. (a) Express $3\cos x° + 5\sin x°$ in the form $k\cos(x° - a°)$ where $k > 0$ and $0 \le a \le 90$.

   4

   (b) Hence solve the equation $3\cos x° + 5\sin x° = 4$ for $0 \le x \le 90$.

   3

5. The graph of the cubic function $y = f(x)$ is shown in the diagram. There are turning points at $(1, 1)$ and $(3, 5)$.

   Sketch the graph of $y = f'(x)$.

   3

   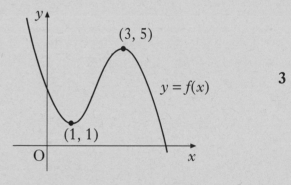

Page three

Marks

6. The circle with centre A has equation $x^2 + y^2 - 12x - 2y + 32 = 0$. The line PT is a tangent to this circle at the point P(5, −1).

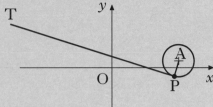

   (a) Show that the equation of this tangent is $x + 2y = 3$.

4

   The circle with centre B has equation $x^2 + y^2 + 10x + 2y + 6 = 0$.

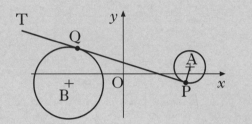

   (b) Show that PT is also a tangent to this circle.

5

   (c) Q is the point of contact. Find the length of PQ.

2

7. An open cuboid measures internally x units by $2x$ units by h units and has an inner surface area of 12 units^2.

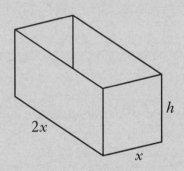

   (a) Show that the volume, V units^3, of the cuboid is given by $V(x) = \frac{2}{3}x(6 - x^2)$.

3

   (b) Find the exact value of x for which this volume is a maximum.

5

8. The amount A_t micrograms of a certain radioactive substance remaining after t years decreases according to the formula $A_t = A_0 e^{-0.002t}$, where A_0 is the amount present initially.

   (a) If 600 micrograms are left after 1000 years, how many micrograms were present initially?

3

   (b) The half-life of a substance is the time taken for the amount to decrease to half of its initial amount. What is the half-life of this substance?

4

Marks

9. An architectural feature of a building is a wall with arched windows. The curved edge of each window is parabolic.

   The second diagram shows one such window. The shaded part represents the glass.

   The top edge of the window is part of the parabola with equation $y = 2x - \frac{1}{2}x^2$.

   Find the area in square metres of the glass in one window.

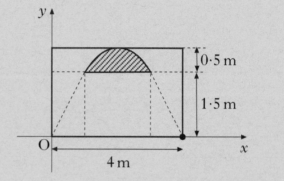

8

[END OF QUESTION PAPER]

[BLANK PAGE]

[BLANK PAGE]

X100/301

NATIONAL QUALIFICATIONS 2008	TUESDAY, 20 MAY 9.00 AM – 10.30 AM	MATHEMATICS HIGHER Paper 1 (Non-calculator)

Read carefully

Calculators may <u>NOT</u> be used in this paper.

Section A – Questions 1–20 (40 marks)

Instructions for completion of **Section A** are given on page two.

For this section of the examination you must use an **HB pencil**.

Section B (30 marks)

1   Full credit will be given only where the solution contains appropriate working.

2   Answers obtained by readings from scale drawings will not receive any credit.

Read carefully

1 Check that the answer sheet provided is for **Mathematics Higher (Section A)**.

2 For this section of the examination you must use an **HB pencil** and, where necessary, an eraser.

3 Check that the answer sheet you have been given has **your name**, **date of birth**, **SCN** (Scottish Candidate Number) and **Centre Name** printed on it.

Do not change any of these details.

4 If any of this information is wrong, tell the Invigilator immediately.

5 If this information is correct, **print** your name and seat number in the boxes provided.

6 The answer to each question is **either** A, B, C or D. Decide what your answer is, then, using your pencil, put a horizontal line in the space provided (see sample question below).

7 There is **only one correct** answer to each question.

8 Rough working should **not** be done on your answer sheet.

9 At the end of the exam, put the **answer sheet for Section A inside the front cover of your answer book**.

Sample Question

A curve has equation $y = x^3 - 4x$.

What is the gradient at the point where $x = 2$?

  A 8

  B 1

  C 0

  D −4

The correct answer is **A**—8. The answer **A** has been clearly marked in **pencil** with a horizontal line (see below).

Changing an answer

If you decide to change your answer, carefully erase your first answer and using your pencil, fill in the answer you want. The answer below has been changed to **D**.

FORMULAE LIST

Circle:

The equation $x^2 + y^2 + 2gx + 2fy + c = 0$ represents a circle centre $(-g, -f)$ and radius $\sqrt{g^2 + f^2 - c}$.

The equation $(x - a)^2 + (y - b)^2 = r^2$ represents a circle centre (a, b) and radius r.

Scalar Product: $\mathbf{a}.\mathbf{b} = |\mathbf{a}|\,|\mathbf{b}| \cos\theta$, where θ is the angle between $\mathbf{a}$ and $\mathbf{b}$

or $\mathbf{a}.\mathbf{b} = a_1b_1 + a_2b_2 + a_3b_3$ where $\mathbf{a} = \begin{pmatrix} a_1 \\ a_2 \\ a_3 \end{pmatrix}$ and $\mathbf{b} = \begin{pmatrix} b_1 \\ b_2 \\ b_3 \end{pmatrix}$.

Trigonometric formulae:
$$\sin(A \pm B) = \sin A \cos B \pm \cos A \sin B$$
$$\cos(A \pm B) = \cos A \cos B \mp \sin A \sin B$$
$$\sin 2A = 2\sin A \cos A$$
$$\cos 2A = \cos^2 A - \sin^2 A$$
$$= 2\cos^2 A - 1$$
$$= 1 - 2\sin^2 A$$

Table of standard derivatives:

$f(x)$	$f'(x)$
$\sin ax$	$a \cos ax$
$\cos ax$	$-a \sin ax$

Table of standard integrals:

$f(x)$	$\int f(x)\,dx$
$\sin ax$	$-\dfrac{1}{a}\cos ax + C$
$\cos ax$	$\dfrac{1}{a}\sin ax + C$

[Turn over

[X100/301]   *Page three*

SECTION A

ALL questions should be attempted.

1. A sequence is defined by the recurrence relation

$$u_{n+1} = 0 \cdot 3u_n + 6 \text{ with } u_{10} = 10.$$

What is the value of u_{12}?

A 6·6

B 7·8

C 8·7

D 9·6

2. The x-axis is a tangent to a circle with centre $(-7, 6)$ as shown in the diagram.

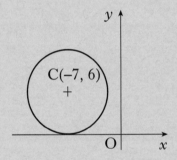

What is the equation of the circle?

A $(x + 7)^2 + (y - 6)^2 = 1$

B $(x + 7)^2 + (y - 6)^2 = 49$

C $(x - 7)^2 + (y + 6)^2 = 36$

D $(x + 7)^2 + (y - 6)^2 = 36$

3. The vectors $\boldsymbol{u} = \begin{pmatrix} k \\ -1 \\ 1 \end{pmatrix}$ and $\boldsymbol{v} = \begin{pmatrix} 0 \\ 4 \\ k \end{pmatrix}$ are perpendicular.

What is the value of k?

A 0

B 3

C 4

D 5

4. A sequence is generated by the recurrence relation $u_{n+1} = 0 \cdot 4 u_n - 240$.

What is the limit of this sequence as $n \to \infty$?

A -800

B -400

C 200

D 400

5. The diagram shows a circle, centre $(2, 5)$ and a tangent drawn at the point $(7, 9)$.

What is the equation of this tangent?

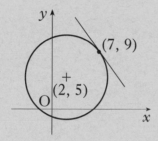

A $y - 9 = -\dfrac{5}{4}(x - 7)$

B $y + 9 = -\dfrac{4}{5}(x + 7)$

C $y - 7 = \dfrac{4}{5}(x - 9)$

D $y + 9 = \dfrac{5}{4}(x + 7)$

[Turn over

6. What is the solution of the equation $2\sin x - \sqrt{3} = 0$ where $\frac{\pi}{2} \le x \le \pi$?

A $\quad \dfrac{\pi}{6}$

B $\quad \dfrac{2\pi}{3}$

C $\quad \dfrac{3\pi}{4}$

D $\quad \dfrac{5\pi}{6}$

7. The diagram shows a line L; the angle between L and the positive direction of the x-axis is $135°$, as shown.

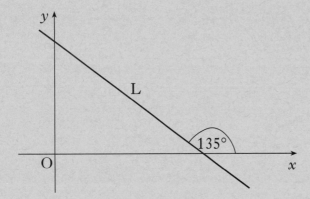

What is the gradient of line L?

A $\quad -\dfrac{1}{2}$

B $\quad -\dfrac{\sqrt{3}}{2}$

C $\quad -1$

D $\quad \dfrac{1}{2}$

8. The diagram shows part of the graph of a function with equation $y = f(x)$.

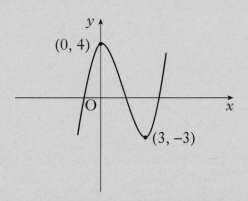

Which of the following diagrams shows the graph with equation $y = -f(x - 2)$?

A

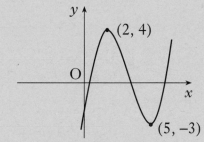

B

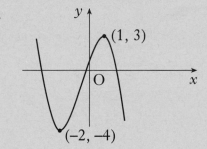

C

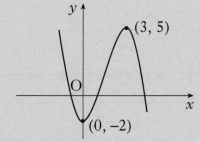

D

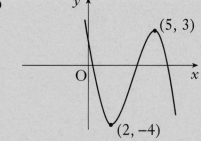

9. Given that $0 \leq a \leq \frac{\pi}{2}$ and $\sin a = \frac{3}{5}$, find an expression for $\sin(x + a)$.

   A $\sin x + \frac{3}{5}$

   B $\frac{4}{5}\sin x + \frac{3}{5}\cos x$

   C $\frac{3}{5}\sin x - \frac{4}{5}\cos x$

   D $\frac{2}{5}\sin x - \frac{3}{5}\cos x$

10. Here are two statements about the roots of the equation $x^2 + x + 1 = 0$:

    (1) the roots are equal;
    (2) the roots are real.

    Which of the following is true?

    A   Neither statement is correct.

    B   Only statement (1) is correct.

    C   Only statement (2) is correct.

    D   Both statements are correct.

11. E(−2, −1, 4), P(1, 5, 7) and F(7, 17, 13) are three collinear points.

    P lies between E and F.

    What is the ratio in which P divides EF?

    A   1:1

    B   1:2

    C   1:4

    D   1:6

12. In the diagram RSTU, VWXY represents a cuboid.

$\overrightarrow{SR}$ represents vector $\boldsymbol{f}$, $\overrightarrow{ST}$ represents vector $\boldsymbol{g}$ and $\overrightarrow{SW}$ represents vector $\boldsymbol{h}$.

Express $\overrightarrow{VT}$ in terms of $\boldsymbol{f}$, $\boldsymbol{g}$ and $\boldsymbol{h}$.

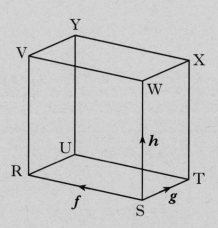

A $\overrightarrow{VT} = \boldsymbol{f} + \boldsymbol{g} + \boldsymbol{h}$

B $\overrightarrow{VT} = \boldsymbol{f} - \boldsymbol{g} + \boldsymbol{h}$

C $\overrightarrow{VT} = -\boldsymbol{f} + \boldsymbol{g} - \boldsymbol{h}$

D $\overrightarrow{VT} = -\boldsymbol{f} - \boldsymbol{g} + \boldsymbol{h}$

13. The diagram shows part of the graph of a quadratic function $y = f(x)$.

The graph has an equation of the form $y = k(x - a)(x - b)$.

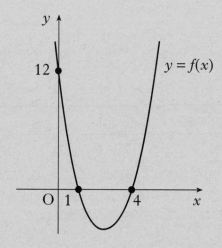

What is the equation of the graph?

A $y = 3(x - 1)(x - 4)$

B $y = 3(x + 1)(x + 4)$

C $y = 12(x - 1)(x - 4)$

D $y = 12(x + 1)(x + 4)$

14. Find $\int 4\sin\,(2x+3)\,dx$.

    A $-4\cos\,(2x+3)+c$

    B $-2\cos\,(2x+3)+c$

    C $4\cos\,(2x+3)+c$

    D $8\cos\,(2x+3)+c$

15. What is the derivative of $(x^3+4)^2$?

    A $(3x^2+4)^2$

    B $\dfrac{1}{3}\,(x^3+4)^3$

    C $6x^2(x^3+4)$

    D $2(3x^2+4)^{-1}$

16. $2x^2+4x+7$ is expressed in the form $2(x+p)^2+q$.

What is the value of q?

    A   5

    B   7

    C   9

    D 11

17. A function f is given by $f(x)=\sqrt{9-x^2}$

What is a suitable domain of f?

    A $x\geq 3$

    B $x\leq 3$

    C $-3\leq x\leq 3$

    D $-9\leq x\leq 9$

18. Vectors $\boldsymbol{p}$ and $\boldsymbol{q}$ are such that $|\boldsymbol{p}| = 3$, $|\boldsymbol{q}| = 4$ and $\boldsymbol{p}.\boldsymbol{q} = 10$.

    Find the value of $\boldsymbol{q}.(\boldsymbol{p} + \boldsymbol{q})$.

    A   0

    B   14

    C   26

    D   28

19. The diagram shows part of the graph whose equation is of the form $y = 2m^x$.

    What is the value of m?

    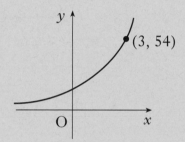

    A   2

    B   3

    C   8

    D   18

20. The diagram shows part of the graph of $y = \log_3(x - 4)$.

    The point $(q, 2)$ lies on the graph.

    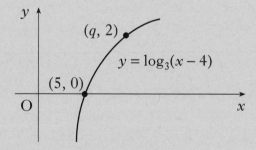

    What is the value of q?

    A   6

    B   7

    C   8

    D   13

    [END OF SECTION A]

SECTION B

ALL questions should be attempted.

Marks

21. A function f is defined on the set of real numbers by $f(x) = x^3 - 3x + 2$.

    (a) Find the coordinates of the stationary points on the curve $y = f(x)$ and determine their nature.   **6**

    (b) (i) Show that $(x - 1)$ is a factor of $x^3 - 3x + 2$.

        (ii) Hence or otherwise factorise $x^3 - 3x + 2$ fully.   **5**

    (c) State the coordinates of the points where the curve with equation $y = f(x)$ meets both the axes and hence sketch the curve.   **4**

22. The diagram shows a sketch of the curve with equation $y = x^3 - 6x^2 + 8x$.

    (a) Find the coordinates of the points on the curve where the gradient of the tangent is -1.   **5**

    (b) The line $y = 4 - x$ is a tangent to this curve at a point A. Find the coordinates of A.   **2**

23. Functions f, g and h are defined on suitable domains by

    $$f(x) = x^2 - x + 10,\ g(x) = 5 - x \text{ and } h(x) = \log_2 x.$$

    (a) Find expressions for $h(f(x))$ and $h(g(x))$.   **3**

    (b) Hence solve $h(f(x)) - h(g(x)) = 3$.   **5**

[END OF SECTION B]

[END OF QUESTION PAPER]

X100/302

| NATIONAL QUALIFICATIONS 2008 | TUESDAY, 20 MAY 10.50 AM – 12.00 NOON | MATHEMATICS HIGHER Paper 2 |

Read Carefully

1 **Calculators may be used in this paper.**

2 Full credit will be given only where the solution contains appropriate working.

3 Answers obtained by readings from scale drawings will not receive any credit.

FORMULAE LIST

Circle:

The equation $x^2 + y^2 + 2gx + 2fy + c = 0$ represents a circle centre $(-g, -f)$ and radius $\sqrt{g^2 + f^2 - c}$.

The equation $(x - a)^2 + (y - b)^2 = r^2$ represents a circle centre (a, b) and radius r.

Scalar Product: $a.b = |a|\,|b|\cos\theta$, where θ is the angle between a and b

or $\quad a.b = a_1b_1 + a_2b_2 + a_3b_3$ where $a = \begin{pmatrix} a_1 \\ a_2 \\ a_3 \end{pmatrix}$ and $b = \begin{pmatrix} b_1 \\ b_2 \\ b_3 \end{pmatrix}$.

Trigonometric formulae:

$$\sin(A \pm B) = \sin A \cos B \pm \cos A \sin B$$
$$\cos(A \pm B) = \cos A \cos B \mp \sin A \sin B$$
$$\sin 2A = 2\sin A \cos A$$
$$\cos 2A = \cos^2 A - \sin^2 A$$
$$= 2\cos^2 A - 1$$
$$= 1 - 2\sin^2 A$$

Table of standard derivatives:

$f(x)$	$f'(x)$
$\sin ax$	$a\cos ax$
$\cos ax$	$-a\sin ax$

Table of standard integrals:

$f(x)$	$\int f(x)\,dx$
$\sin ax$	$-\dfrac{1}{a}\cos ax + C$
$\cos ax$	$\dfrac{1}{a}\sin ax + C$

ALL questions should be attempted.

Marks

1. The vertices of triangle ABC are A(7, 9), B(−3, −1) and C(5, −5) as shown in the diagram.

   The broken line represents the perpendicular bisector of BC.

   (a) Show that the equation of the perpendicular bisector of BC is $y = 2x - 5$.

   4

   (b) Find the equation of the median from C.

   3

   (c) Find the coordinates of the point of intersection of the perpendicular bisector of BC and the median from C.

   3

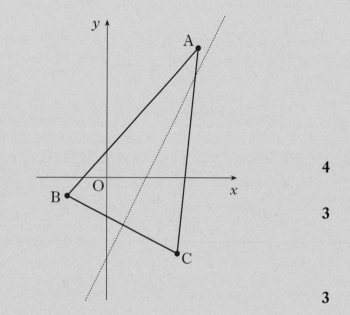

2. The diagram shows a cuboid OABC, DEFG.

   F is the point (8, 4, 6).

   P divides AE in the ratio 2:1.

   Q is the midpoint of CG.

   (a) State the coordinates of P and Q.

   2

   (b) Write down the components of $\overrightarrow{PQ}$ and $\overrightarrow{PA}$.

   2

   (c) Find the size of angle QPA.

   5

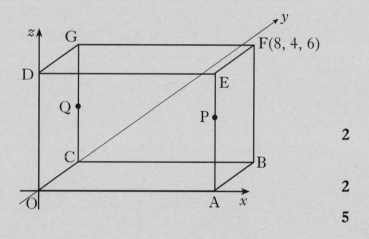

[Turn over

Marks

3. (a) (i) Diagram 1 shows part of the
       graph of $y = f(x)$, where
       $f(x) = p\cos x$.

       Write down the value of p.

       Diagram 1

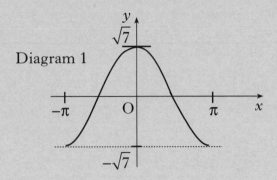

    (ii) Diagram 2 shows part of the
        graph of $y = g(x)$, where
        $g(x) = q\sin x$.

        Write down the value of q.   Diagram 2   2

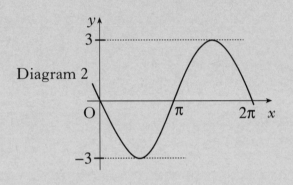

   (b) Write $f(x) + g(x)$ in the form $k\cos(x + a)$ where $k > 0$ and $0 < a < \dfrac{\pi}{2}$.   4

   (c) Hence find $f'(x) + g'(x)$ as a single trigonometric expression.   2

4. (a) Write down the centre and calculate the radius of the circle with equation
       $x^2 + y^2 + 8x + 4y - 38 = 0$.   2

   (b) A second circle has equation $(x - 4)^2 + (y - 6)^2 = 26$.

       Find the distance between the centres of these two circles and hence show
       that the circles intersect.   4

   (c) The line with equation $y = 4 - x$ is a common chord passing through the
       points of intersection of the two circles.

       Find the coordinates of the points of intersection of the two circles.   5

5. Solve the equation $\cos 2x° + 2\sin x° = \sin^2 x°$ in the interval $0 \leq x < 360$.   5

Marks

6. In the diagram, Q lies on the line joining (0, 6) and (3, 0).

OPQR is a rectangle, where P and R lie on the axes and OR = t.

(a) Show that QR = $6 - 2t$.

3

(b) Find the coordinates of Q for which the rectangle has a maximum area.

6

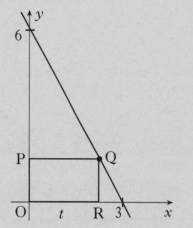

7. The parabola shown in the diagram has equation

$$y = 32 - 2x^2.$$

The shaded area lies between the lines $y = 14$ and $y = 24$.

Calculate the shaded area.

8

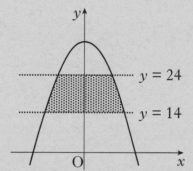

[END OF QUESTION PAPER]

[BLANK PAGE]